TENTH EDITION

ON THE INFLUENCE

OF THE

BLUE COLOR OF THE SKY

IN

Developing Animal and Vegetable Life,

AS ILLUSTRATED IN THE EXPERIMENTS OF

GEN. A. J. PLEASONTON,

Between the Years 1861 and 1871,

AT PHILADELPHIA.

READ BEFORE THE PHILADELPHIA SOCIETY FOR PROMOTING AGRICULTURE,

ON WEDNESDAY, MAY 3, 1871.

PHILADELPHIA:

Inquirer Book and Job Print, 304 Chestnut Street.

1871.

ON THE INFLUENCE

OF THE

BLUE COLOR OF THE SKY

IN

Developing Animal and Vegetable Life,

AS ILLUSTRATED IN THE EXPERIMENTS OF

GEN: A. J. PLEASONTON,

Between the Years 1861 and 1871,

AT PHILADELPHIA.

READ BEFORE THE PHILADELPHIA SOCIETY FOR PROMOTING AGRICULTURE,

ON WEDNESDAY, MAY 3, 1871.

PHILADELPHIA:

INQUIRER BOOK AND JOB PRINT, 304 CHESTNUT STREET.

1871.

The following memoir was read by GEN. A. J. PLEASONTON, before the Philadelphia Society for Promoting Agriculture, on Wednesday, the 3d of May, 1871, at their room, S. W. corner of 9th and Walnut Streets, in the City of Philadelphia, upon the following request:

1309 WALNUT ST., *April 27th, 1871.*

MR DEAR GENERAL:

Will it suit you, and will you do us the favor to explain your process of using glass in improving stock to the Philadelphia Society for Promoting Agriculture, on Wednesday next, the 3d of May, at eleven o'clock, A. M., at their Room, S. W. corner of Ninth and Walnut Streets, (entrance on Ninth street)? You were kind enough to express to me. in conversation, your willingness to give us the result of your experiments.

Yours, very truly,

W. H. DRAYTON,

President.

GENERAL PLEASONTON.

PREFACE.

Having been much interested in the phenomena of the physics of the earth, the author, in offering to his readers a second edition of his work, "On the Influence of the Blue Color of the Sky in Developing Animal and Vegetable Life," may be indulged in his introduction into this preface of some views that his observations have led him to entertain relative to the variations of temperature, and changes of our seasons, which are in harmony with the subjects treated by him in this work.

The first edition of the following memoir was printed for distribution among scientific and literary institutions, and among persons of culture, for the purpose of attracting the attention of those for whom it was intended, to the subjects of which it treats. It was hoped that its publication would invite investigation into the nature, composition, and influences of those great forces which, in the poverty of our language, we call imponderables, that is to say, not to be weighed in the balance, and consequently never to be found wanting. This expectation is likely to be realized, if we may judge from the general interest that appears to be taken in the memoir, which has been manifested in the numerous applications that have been made to the author, from various parts of our country, for copies of it. The edition has now been distributed, yet so many persons who have applied for copies of the memoir are still without it, that it has been deemed advisable to issue another edition.

If, by a course of study, and observation of the great forces of nature, as they are exhibited, not in the laboratory, upon the minutest scale, but in those grand operations by which physical changes are at every moment developed before our eyes, we can succeed in penetrating the mysteries of their origin, of their evolution, of their application, and of their reciprocal conversions into each other, we shall become indeed wise in our generation, and mankind in the future will be able to rejoice in a development never yet reached in any preceding age.

By way of illustration of this idea, we may suggest that this planet is surrounded, at variable altitudes above its surface, by a canopy of cold, increasing in intensity with its distance above the earth. Now, we may ask, what produces the changes of our seasons? We answer, simply the descent or ascent of columns of this canopy of cold!

It has been observed, for any years, that the first frost of the autumn appears in Texas or Louisiana, or some other of the Gulf States, while at the same time no frost is observable in other localities situated much farther to the north—the commonly supposed place of departure of our winters. This frost, therefore, must come from the descent of the cold of the higher atmosphere immediately over the locality where it prevails. Following the valley of the Mississippi and those of its tributaries, frost appears successively in various places along those routes, till it reaches the vallies of the Northern Lakes, running along which it is felt in Northern New York and the New England States, and subsequently in the Middle and Southern Atlantic States. It does not reach the vicinity of Philadelphia until some fifteen or twenty days after it has shown itself on the Gulf of Mexico. Now would it not seem that the influences producing this frost are telluric, and not exclusively solar, as hitherto they have been supposed to be?

We know that in the ocean there are columns of fresh water which differ in temperature from the surrounding sea water, and with which they do not mingle for a long time. So is it for a hundred or more miles at sea, distant from the mouths of the great rivers Amazon, Orinoco, Mississippi, etc., whose fresh waters do not mix with the salt waters surrounding them, owing to the difference of their densities. In like manner the cold air of the upper atmosphere descends in columns of various extent over particular localities, to vary the temperature and change the seasons, on the surface of our earth, without mixing with the warmer and more expanded air beneath, which it displaces.

The spring and summer seasons are produced by increased radiations from the interior heat of the earth, forcing upwards the dense cold of winter, whose particles are so close together as to prevent the intrusion among them of the expanded warm air in its ascent. Much of the heat of the lower atmosphere is also developed in the conversion of vapor into clouds by condensation from cold.

It is in this way that our seasons are changed. Let our savans discover how and why these effects are produced. Until they do, it may be suggested that they are owing to electrical atmospherical disturbances in the upper atmosphere, repelling the negative electricity of those regions, and forcing the cold air to the surface of the earth, where it displaces the warmer and more rarefied and expanded air, and condenses in rain, snow and hail, the vapors it contains, driving the displaced warmer air to the tropics, and the heat from the tropics attracted to the condensed vapor in the clouds in the temperate zones to liquefy them in rain, producing winter.

In the opposite manner the warm seasons of spring and summer are produced by the positive electricity of the surface-air of the earth becoming warmed by increased radiation of heat from the interior of the earth, repelling itself, and displacing the upper strata of cold air, till by induction of electricity the temperature of the season is established.

Geologists tell us that in the early existence of this planet, the greater part of the earth's surface was covered with ice, and that this period of time is called the Glacial Period.

Let us imagine that the igneous action of the elementary substances of the interior of the earth's crust, just before that period, might have been so intense as by the radiation of its heat to the surface of the earth to rarefy the lower atmosphere, converting into vapor the water it contained, and forcing it upward till the whole surface of the earth was almost incandescent.

To restore the equilibrium, the canopy of cold repelled by its own negative electricity from above, which has been increased by the currents of polar electricity, largely developed by this central and interior igneous action—and attracted by the positive electricity in the heated atmosphere below—descended to the surface of the earth, condensing the vapors of the atmosphere into rain, and afterwards into hail and snow, driving the remainder of the warmer air of what we call, now, the temperate zones, to the tropics, and covering the surfaces of the earth, from the poles to the tropics, with a dense mantle of ice, freezing the rivers, bays, and seas of those latitudes. The internal central fires thus concentrated, in due season increased their radiation of heat, and melted the superjacent ice, which, breaking from the sides of glaciers in large masses, slid and rolled to the ocean, there becoming icebergs, and carrying with them those immense boulders which, torn from the mountain sides by the adhesion of the ice, have left the traces of their furrows on the slopes of the mountains, and have marked their courses till, by the melting of the bergs, they have been dropped in the ocean, which subsequently, by its subsidence, have left them dry on the land. If such was the cause of the glacial period, it would require no great stretch of fancy to comprehend the deluge of Deucalion or that of our great ancestor Noah, when the rain descended for forty days;

occasioned no doubt by a lesser descent of the canopy of cold (limiting its effect to the condensation of the vapors of the atmosphere into rain) than that which produced the glacial period.

If such effects follow from such causes, we need not be at a loss to account for the changes of our seasons, or the daily variations of temperature in every locality.

This edition of our memoir has been printed upon tinted paper with blue ink, as an experiment, in an attempt to relieve the eyes of the reader from the great glare, occasioned by the reflection of gas light at night from the white paper usually employed in the printing of books. If it shall succeed we may hope to see the tinted paper introduced for all books and periodicals.

PHILADELPHIA, July 29, 1871.

Mr. President and Gentlemen of The Philadelphia Society for Promoting Agriculture.

At the request of my old friend and your respected President, I have attended your meeting this morning to impart to you the results of certain experiments that I have made within the last ten years in attempts to utilize the blue color of the sky in the development of vegetable and animal life.

I may premise that for a long time I have thought that the blue color of the sky, so permanent and so all-pervading, and yet so varying in intensity of color, according to season and latitude, must have some abiding relation and intimate connection with the living organisms on this planet.

Deeply impressed with this idea, in the autumn of the year 1860, I commenced the erection of a cold grapery on my farm in the western part of this city. I remembered that while a student of chemistry I was taught that in the analysis of the ray of the sun by the prism, in the year 1666, by Sir Isaac Newton, he had resolved it into the seven primary rays, viz: red, orange, yellow, green, blue, indigo and violet, and had discovered that these elementary rays had different indices of refraction; that for the *red* ray at one side of the solar spectrum being the least, while that of the *violet* at the opposite side thereof was the greatest, from which he deduced his celebrated doctrine of *the different refrangibility of the rays of light;* and further, that Sir John Herschel in his subsequent investigation of the properties of light had shown that the chemical power of the solar ray is greatest in the *blue rays,* which give the least light of any of the *luminous* prismatic radiations, but the largest quantity of solar heat, and that later experiments established the fact of the stimulating influence of the blue rays upon vegetation. Having concluded to make a practical application of the properties of the blue and violet rays of light just referred to in stimulating vegetable life, I began to inquire in every accessible direction if this stimulating quality of the blue or violet ray had ever received any practical useful application. My inquiries developed the facts that various experiments had been made in England and on the European continent with glass colored with each of the several primary rays, but that they were so unsatisfactory in their results that nothing useful came of them so far as any improvement in the process of developing vegetation was concerned. Finding no beaten track, I was left to grope my way as best I could under the guidance of the violet ray alone. My grapery was finished in March, 1861. Its dimensions were, 84 feet long, 26 feet wide, 16 feet high at the ridge, with a double-pitched roof. It was built at the foot of a terraced garden, in the direction of N. E. by E. to S. W. by W. On three sides of it there was a border 12 feet wide, and on the fourth or N. E. by E. side the border was only five feet wide, being a walk of the garden. The borders inside and outside were excavated 3 feet 6 inches deep, and were filled up with the usual nutritive matter, carefully prepared for growing vines. I do not think they differed essentially from thousands of other borders which have been made in many parts of the world. The first question to be solved on the completion of the frame of the grapery, was the proportion of blue or violet glass to be used on the roof. Should too much be used, it would reduce the temperature too much, and cause a failure of the experiment; if too little, it would not afford a fair test. At a venture I adopted every eighth row of glass on the roof to be violet colored, alternating the rows on opposite sides of the roof, so that the sun in its daily course should cast a beam of violet light on every leaf in the grapery. Cuttings of vines of some twenty varieties of grapes, each one year old, of the thickness of a pipe-stem, and cut close to the pots containing them, were planted in the borders inside and outside of the grapery, in the early part of April, 1861. Soon after being planted the growth of the vines began. Those on the outside were trained through earthen pipes in the walls to the inside, and as they grew they were tied up to the wires like those which had been planted within. Very soon the vines began to attract great notice of all who saw them from the rapid growth they were making. Every day disclosed some new extension, and the gardener was kept busy in tying up the new wood which the day before he had not observed. In a few weeks after the vines had been planted, the walls and inside of the roof were closely covered with the most luxurious and healthy development of foliage and wood.

In the early part of September, 1861, Mr. Robert Buist, Sr., a noted seedsman and distinguished horticulturist from whom I had procured the vines, having heard of their wonderful growth, visited the grapery. On entering it he seemed to be lost in amazement at what he saw; after examining it very carefully, turning to me, he said, "General! I have been cultivating plants and vines of various kinds for the last forty years; I have seen some of the best vineries and conservatories in England and Scotland, but I have never seen anything like this growth." He then measured some of the vines and found them forty-five feet in length, and an inch in diameter at the distance of one foot above the ground; and these dimensions were the growth of only five months! He then remarked, "I visited last week a new grapery near Darby, the vines in which I furnished at the same time I did yours; they were of the same varieties, of like age and size, when they were planted as yours; they were planted at the same time with yours. When I saw them last week, they were puny spindling plants not more than five feet long, and scarcely increased in diameter since they were planted—and yet they have had the best possible care and attendance!"

The vines continued healthy and to grow, making an abundance of young wood during the remainder of the season of 1861.

In March of 1862 they were started to grow, having been pruned and cleaned in January of that year. The growth in this second season was, if anything, more remarkable than it had been in the previous year. Besides the formation of new wood and the display of the most luxuiriant foliage, there was a wonderful number of bunches of grapes, which soon assumed the most remarkable proportions—the bunches being of extraordinary magnitude, and the grapes of unusual size and development.

In September of 1862 the same gentleman Mr. Robert Buist, Sr., who had visited the grapery the year before came again—this time accompa nied by his foreman. The grapes were then beginning to color and to ripen rapidly. On entering the grapery, astonished at the wonderful display of foliage and fruit which it presented, he stood for a while in silent amazement; he then slowly walked around the grapery several times, critically examining its wonders; when taking from his pocket paper and pencil, he noted on the paper each bunch of grapes, and estimated its weight, after which aggregating the whole, he came to me and said, "General! do you know that you have 1200 pounds of grapes in this grapery?" On my saying that I had no idea of the quantity it contained, he continued, "you have indeed that weight of fruit, but I would not dare to publish it, for no one would believe me." We may well conceive of his astonishment at this product when we are reminded that in grape-growing countries where grapes have been grown for centuries, that a period of time of from five to six years will elapse before a single bunch of grapes can be produced from a young vine—while before him in the second year of the growth of vines which he himself had furnished only seventeen months before, he saw this remarkable yield of the finest and choicest varieties of grapes. He might well say that an account of it would be incredible.

During the next season (1863) the vines again fruited and matured a crop of grapes estimated by comparison with the yield of the previous year to weigh about two tons; the vines were perfectly healthy and free from the usual maladies which affect the grape. By this time the grapery and its products had become partially known among cultivators, who said that such excessive crops would exhaust the vines, and that the following year there would be no fruit, as it was well known that all plants required rest after yielding large crops; notwithstanding, new wood was formed this year for the next year's crop, which turned out to be quite as large as it had been in the season of 1863, and so on year by year the vines have continued to bear large crops of fine fruit without intermission for the last nine years. They are now healthy and strong, and as yet show no signs of decrepitude or exhaustion.

The success of the grapery induced me to make an experiment with animal life. In the autumn of 1869 I built a piggery and introduced into the roof and three sides of it violet-colored and white glass in equal proportions—half of each kind. Separating a recent litter of Chester county pigs into two parties, I placed three sows and one barrow pig in the ordinary pen, and three other sows and one other barrow pig in the pen under the violet glass. The pigs were all about two months old. The weight of the pigs was as follows, viz: Under the violet glass, No. 1 sow, 42 lbs., No. 2, a barrow pig, 45½ lbs., No. 3, a sow, 38 lbs., No. 4, a sow 42, lbs., their aggregate weight 167½ lbs. The weight of the others in the common pen was as follows, viz: No. 1., a sow, 50 lbs., No. 2, a sow, 48 lbs., No. 3, a barrow big, 59 lbs., No. 4, a sow, 46 lbs; their aggregate weight was 203 lbs. It will be observed that each of the pigs under the violet glass was lighter in weight than the lightest in weight pig of those under the sunlight alone in the common pen. The two sets of pigs were treated exactly alike; fed with the same kinds of food at equal intervals of time, and with equal quantities by measure at each meal, and were attended by the same man. They were put in the pens on the 3d day of November, 1869, and kept there until the 4th day of March, 1870, when they were weighed again. By some misconception of my orders, the separate weight of each pig was not had. The aggregate weight of the three sows under the violet light on the 3d of November, 1869, was 122 lbs; on the 4th of March, 1870, it was 520 lbs., increase 398 lbs.

The aggregate weight of the three sows in the old pens on the 3d of November, 1869, was 144 lbs., and on the 4th of March, 1870, it was 530 lbs., increase 386 lbs., or 12 lbs. less than those under the violet glass had gained.

The weight of the barrow pig in the common pen on the 3d of November, 1869, was 59 lbs., and on the 4th of March, 1870, it was 210 lbs., increase 151 lbs. The weight of the barrow pig under the violet light, on the 3d of November, 1869, was 45½ lbs., and on the 4th of March, 1870, it was 170 lbs., increase 124½ lbs. The large increase of the weight of the barrow pig in the common pen is to be attributed to his superior size and weight on being put in the same common pen with the three sows, and which enabled him to seize upon and appropriate to himself more than his share of the common food.

If the barrow pig under the violet light had increased at the rate of increase of the barrow pig in the common pen, his weight on the 4th March, 1870, would have been only $161\frac{84}{100}$ lbs. instead of his actual weight of 170 lbs.—showing his rate of increase of weight to have been $8\frac{36}{100}$ lbs. more than that of the other barrow pig.

If the barrow pig under the sunshine in the common pen had increased at the rate of increase of the barrow pig under the violet glass, his weight on the 4th of March, 1870, should have been $224\frac{42}{100}$ lbs. instead of 210 lbs., his actual weight at that date.

By these comparisons it seems obvious that the influence of the violet-colored glass was very marked, although it must be borne in mind that owing to the great declination of the sun during the period of the experiment and the consequent comparative feebleness of the force of the actinic or chemical rays of the blue sky at that time, the effect was not so great as it would have been at a later period of the season; but the time of the experiment was selected for that very reason. The animals were not fed to produce fat or increase of size, but simply to ascertain, if practicable, whether by the ordinary mode of feeding usual on farms in this country, the development of stock could be hastened by exposing them in pens to the combined influence of sunlight and the transmitted rays of the blue sky.

My next experiment was with an Alderney bull calf born on the 26th of January, 1870; at its birth it was so puny and feeble that the man who attends upon my stock, a very experienced hand, told me that it could not live. I directed him to put it in one of the pens under the violet glass. It was done. In 24 hours a very sensible change had occurred in the animal. It had arisen on its feet, walked about the pen, took its food freely by the finger, and manifested great vivacity. In a few days its feeble condition had entirely disappeared. It began to grow, and its development was marvelous. On the 31st March, 1870, 2 months and 5 days after its birth, its rapid growth was so apparent, that as its hind quarter was then growing, I told my son to measure its height, and to note down in writing the height of the hind quarter, and the time of measurement—which he did. On the 20th of the following May (1870), just fifty days afterwards, my son again measured the hind quarter, and found that in that time it had gained *exactly six inches in height, carrying its lateral development with it.* Believing the question solved, the calf was turned into the barn-yard, and when mingling with the cows he manifested every symptom of full masculine vigor, though at the time he was only four months old. Since the 1st of April of this year, when he was just 14 months old, he has been kept with my herd of cows, and has fulfilled every expectation that I had formed of him. He is now one of the best developed animals that can be found any where.

These, gentlemen, are the experiments about which your curiosity has been excited. If by the combination of sunlight and blue light from the sky, you can mature quadrupeds in twelve months with no greater supply of food than would be used for an immature animal in the same period, you can scarcely conceive of the immeasurable value of this discovery to an agricultural people. You would no longer have to wait five years for the maturity of a colt; and all your animals could be produced in the greatest abundance and variety. A prominent member of the bar a short time since told me that his sister, who is a widow of a late distinguished general in the army, had applied blue light to the rearing of poultry, with the most remarkable success, after having heard of my experiments. In regard to the human family, its influence would be wide spread—you could not only in the temperate regions produce the early maturity of the tropics, but you could invigorate the constitutions of invalids, and develop in the young, a generation, physically and intellectually, which might become a marvel to mankind. Architects would be required to so arrange the introduction of these mixed rays of light into our houses, that the occupants might derive the greatest benefit from their influence. Mankind will then not only be able to live fast, but they can live well and also live long.

Let us attempt an explanation of this phenomenon. It is well known that differences of temperature evolve electricity, as do also evaporation, pressure suddenly produced or suddenly removed, in which may be comprised a blow or stroke, as, for instance, from the horseshoe in the rapid motion of a horse on a stone in the pavement, striking fire, which is kindled by the electricity evolved in the impact, or, again, from the collision of two silicious stones in which there is no iron, is electricity produced.

Friction even of two pieces of dried wood excites combustion by the evolution of hydrogen gas which bursts into flame when brought into contact with the opposite electricity evolved by the heat. Chrystallization, the freezing of water, the melting of ice or snow—every act of combination in respiration, every movement and contraction of organic tissues, and, indeed, every change in the form of matter evolve electricity, which in turn contributes to form new modifications of the matter which has yielded it.

The diamond, about whose origin so much mystery has always existed, it is likely, is the product of the decomposition of carbonic acid gas in the higher atmosphere by electricity, liberating the oxygen gas, converting it into ozone, fusing the carbon, and by the intense cold there prevailing, which is of opposite electricity, chrystallizing the fused carbon, which is precipitated by its gravity to the earth.

To the repellent affinity of electricity are we indebted for the expansive force of steam whose power wields the mighty trip hammer, propels the ship through the ocean, and draws the train over the land—and to the opposite electricities of the heated steam and the cold water introduced into the boiler to replenish it, do we owe those terrible explosions in steam boilers whose prevention has hitherto defied human skill. But the most interesting application of electricity, is in nature's development of vegetation. Let us illustrate it:

Seed perfectly dried, but still retaining the vital principle, like the seed of wheat preserved for thousands of years in mummy cases in the catacombs of Egypt, if planted in a soil of the richest alluvial deposits, also thoroughly dried, will not germinate. Why? Let us examine. The alluvial soil is composed of the *debris* of hills and mountains containing an extensive variety of metallic and metalloid compounds mingled with the remains of vegetable and animal matter in a state of great comminution, washed by the rains and carried by freshets into the depressions of the surface of the earth. These various elements of the soil have different electrical attributes. In a perfectly dry state no electrical action will occur among them, but let the rain, bringing with it ammonia and carbonic acid, in however minute quantities, from the upper atmosphere, fall upon this alluvial soil, so as to moisten its mass within the influence of light, heat, and air, and plant your seed within it, and what will you observe? Rapid germination of the seed. Why? The slightly acidulated, or it may be alkaline water of the rain has formed the medium to excite galvanic currents of electricity in the heterogeneous matter of the alluvial soil—the vitality of the seed is developed and vegetable life is the result. Hence vegetable life owes its existence to electricity. Herein consists the secret of successful agriculture. If you can maintain the currents of electricity at the roots of plants by supplying the acidulated or alkaline moisture to excite them during droughts, you will secure the most abundant and unvarying crops. To do this, your soil should be composed of the most varied elements, mineral, earthy, alkaline, vegetable, and animal matter in a state of greatly comminuted decomposition.

The poverty of soils arises from the homogeneous character of their composition. A soil altogether clayey, or composed of silicious sand, or the *debris* of limestone, or of alkaline substances exclusively, must necessarily be barren for the want of electrical excitement, which no one of the said elements will produce; but commingle them all with the addition of decomposed vegetable and animal matter, and you will form a soil which will amply reward the toil of the husbandman.

What do you suppose has produced the giant trees of California? Electricity! Since the west coast of America has been known to Europeans, and perhaps for centuries before, it has been subjected to the most devastating earthquakes. From the Straits of Magellan to the Arctic Ocean, traces of volcanic action are everywhere visible. Its mountains have been upheaved, broken, torn asunder, and sometimes, like Ossa upon Pelion, one has been superimposed on another.

All volcanic countries are noted in the temperate regions, where they produce anything, for the exuberance of their vegetable productions. Etna has been famous for its large Chestnut trees, which have given a name Catania to the town near its base.

The mineral richness of California has doubtless, by the *debris* of its mountains, carried into the valleys where grow these large trees, furnished an immense deposite of various matter which, under the favorable circumstances of the localities, have maintained for ages a healthful electrical excitement resulting through centuries of undisturbed growth in these vegetable wonders.

Who is there that has not been struck with admiration in looking upon the firmament, when the atmosphere was clearest, and was unclouded by the slightest vapor,—when, in the brightness of sunlight, it would put on its livery of blue, and display its resplendent and glorious beauties? How many myriads of mankind, in all ages, have gazed upon this magnificent arch, of what men call "sky;" and how few have ever asked the question, Why is the sky blue? and why should its intensity of blue vary in different latitudes, and in different seasons?

Humboldt said he had never seen its blue so intense as in the tropics and under the equator. Arctic navigators have also declared, that in the arctic regions the intensity of the blue color of the sky was amazing. Here are two extremes of latitude displaying the same effect; and in our own temperate region many have observed a variation in the intensity of the blue of the sky, in different seasons, extending from the early spring until the close of autumn, but never equaling in depth of color what is represented of it, either in the tropics or in the arctic or antarctic regions.

On no part of our planet is the development of vegetable life so grand, so various, so excessive and so constant as in the tropics and in the equatorial regions. While this wonderful display of vegetation is observed in these regions, the exuberance of animal life and the rapid growth of vegetable life in the arctic regions are said to be unequaled in any other part of our world. Let us see if these results in the two natural kingdoms may not be attributed largely to the same cause.

Recent discoveries have shown that the Zodiacal light over the equator and the auroræ borealis and australis are evolutions of electricity. In the arctic regions there is little doubt that the auroræ are constantly evolved, though they are not always visible. They have been seen to emerge from the surface of the ocean, at short distances from the observers, and ascending into the upper regions of the atmosphere, to present those corruscations of brilliant light, shooting as it were to the equatorial regions, in rapid flashes, for which they have been noted wherever observed.

In the equatorial regions it is well known that at certain periods of the year the accumulation of electricity in the upper atmosphere is so excessive, that the earth is shaken with thunderbolts, and the air illuminated by day as well as night with constant sheets of electric flame, as they rush with frightful velocity to their great centre of attraction, the earth and ocean in those regions. Whence does this electricity come, and where does it go?

If we may be permitted to form a conjecture, we might suggest that the sixty odd primary elements which enter into the composition of the crust of our planet—such as carbon, sulphur, phosphorus, oxygen, nitrogen, hydrogen, the metals, the metalloids, etc.—having been endowed by the Creator with different electrical qualities and conditions—when they were assembled together in this planet, evolved in the interior thereof electricity, light, heat, and magnetism in certain or variable qualities and quantities. These constitute the forces which in all probability cause the rotation of the earth upon its axis, and assist in its revolution around the sun. The electricity of the interior of the earth is supposed to be positive electricity —which, as soon as evolved there, would be repelled according to the law of electricity of the same character repelling itself —towards the poles of the earth, and escaping there, would be attracted by the negative electricity which surrounds the upper atmosphere, and would display itself by night as auroræ, corruscating toward the equator, to be there attracted by the heated equatorial regions, and descending to the earth, to be again absorbed by it, for further use. This escape of polar electricity into the upper atmosphere, and forming at night the auroræ, when visible, and by day the blue firmament or sky, will account for the intensity of the blue color of the sky both in the arctic regions and the equatorial regions.

This positive electricity of the central interior of the earth, repelling itself towards the poles, and from there into the atmosphere through the arctic and antarctic oceans, and attracted there by the negative electricity of the upper atmosphere, forms, by the union of the two electricities, the auroras, causing those crackling detonations heard during the prevalence of the most brilliant auroras, in high latitudes and evolving light, which, seen through the vaporous atmosphere of those latitudes, is displayed by refractions of its rays in the luminous corruscations of varying tints as the rays of the sun or moon are converted into the tints of the rainbow.

The negative electricity of those frigid regions attracted to the equator through the upper atmosphere is there concentrated in enormous quantities, which are conducted and discharged into the earth or ocean in the tropics, by the incessant fall of water in rain during the rainy seasons, every drop of water being a conductor of electricity, and every leaf of vegetation assisting in the conduct and distribution of this wonderful force into the earth.

As under certain circumstances electricity becomes magnetism, and this again is converted into electricity, we can comprehend how the auroral rays in some instances, following the law of dia-magnetism, are attracted in the northern hemisphere towards the southwest—magnetic currents flowing from east to west in opposition to the earth's motion from west to east; hence in the auroras you have rays shooting to the zenith over the equator, and others moving southwest, and others again due west.

The simultaneous appearance of auroras frequently observed in opposite hemispheres in corresponding latitudes would go to show their origin from a common impulse in the central interior repelling them towards the poles from under the equator.

We now come to a presumed explanation of one of the reasons for the blue color of the sky.

The sun's ray, or what is called the white light of the sun, was resolved by means of a glass prism, by Sir Isaac Newton, into the seven primary rays of light, viz., red, blue, violet, etc., and their combination again produced the white light– showing both synthetically and analytically of what the sun's light was composed.

It was announced in England about the beginning of this century, that the red ray of light was heating, the yellow ray was illuminating, *and the blue ray in a remarkable degree stimulated the development of vegetable life.*

From this discovery we can imagine the immense influence which the intensely blue color of the sky in the equatorial regions has and always has had in conjunction with the sun's white light, and the heat and moisture of those regions, upon the development there of vegetable life.

This intensely blue color of the sky in the arctic regions may also serve to explain the exuberance of animal life there. It being known that the deeper water of the arctic ocean is much warmer than the surface water which is often frozen, furnishes abundant food for its inhabitants. The increased temperature of this deep water is probably derived from radiation of heat from the interior of the earth under it—as all those regions are more or less volcanic; witness Iceland, Jean Mayer, Spitsbergen, etc. The laws of animal and vegetable life being very analogous, what would stimulate one would probably have a similar effect upon the other.

In the arctic waters you have warmth, food, light and electricity, escaping through the waters into the air, and all stimulating life.

Whoever has noticed the color of the electric spark in atmospheric air, from an electrical machine, will readily recognize its likeness of color to the blue color of the sky.

If experiments should be instituted to ascertain the electrical condition of the sky, *as associated with its blue color*, and they should satisfactorily establish the connection, the result would prove to be one of the greatest blessings ever conferred upon mankind. What strength of vitality could be infused into the feeble young, the mature invalid, and the decrepit octogenarian! How rapidly might the various races of our domestic animals be multiplied, and how much might their individual proportions be enlarged!

One of the most beautiful illustrations of the mighty influence of the blue color of the sky upon vegetation, is to be found in the green color of the leaves of plants. It is known that blue and yellow when mixed produce green, which is darker when the blue is in excess over the yellow, and the reverse when the yellow predominates. Now let us observe the process of germination. Seeds are planted in the soil—at first a white worm-like thread at the lower part of the seed appears; it is white, and contains all the primary rays of light; it is the root of the plant, and remaining under the soil continues white. At the upper end of the seed also appears a white swelling, which continues to grow upward till it approaches the surface of the soil, when a change occurs in its color. This is the leaf; it absorbs yellow from the soil which is brown (composed of yellow and black), and as it comes within the influence of the blue sky, it absorbs from it the blue light, which mixing with the yellow already absorbed, produces at first a yellowish-green, which finally assumes the deeper tinge of green that is natural to the plant. The plant blossoms, forms its seeds and seed-vessels, and having fulfilled its mission, the blue color of the leaves is eliminated, the leaves become yellow, and absorbing the carbon of the plant, they change their color to brown; the sap-vessels of the leaves are choked by the carbon; the leaves are dead and fall to the ground. Thus the blue ray is the symbol of vitality—the yellow ray that of decay and death.

Robert Hunt, in his Researches on Light, says "that the rays of greatest refrangibility, viz., the violet, &c., favor dis-oxygenation, but the rays of least refrangibility, viz., red, orange, &c., favor oxygenation."

"The experiments of Seunebier show that the most refrangible of the solar rays, viz., the violet, are the most active in determining the decomposition of carbonic acid gas by plants."

These experiments have been confirmed by Mr. Robert Hunt, who says, "that experiments have been made with absorbent media, and the light which has been carefully analyzed, permeating under the influence of *blue light*, in every instance oxygen gas has been collected, but not any under the energetic action of yellow or red light. * * It is only the green parts of plants which absorb carbonic acid: the flowers absorb oxygen gas. Plants grow in soils composed of divers materials, and they derive from these by the soluble powers of water, which is taken up by the roots, and by mechanical forces carried over every part, carbonic acid, carbonates and organic matters containing carbon. Evaporation is continually going on, and this water escapes freely from the leaves during the night when the functions of the vegetable, like those of the animal world, are at rest, and carries with it carbonic acid. Water and carbonic acid are sucked up by capillary attraction, and both evaporate from the exterir part of fhe leaves."

"There is no reversion of the processes which are necessary to support the life of a plant. The samc functions are operating in the same way by day and by night, but differing greatly in degree. During the hours of sunshine the whole of the carbonic acid absorbed by the leaves or taken up with water by the roots is decomposed, all the functions of the plant are excited, the processes of inhalation and exhalation are quickened, and the plant pours out to the atmosphere streams of pure oxygen at the same time as it removes a large quantity of deleterious carbonic acid from it. In the shade the exciting power being lessened, these operations are slower, and in the dark they are very nearly, but certainly not quite, suspended."

"Although a blue glass or fluid may appear to absorb all the rays except the most refrangible ones, which have usually been considered as the least calorific of the solar rays; *yet it is certain that some principle* has permeated the glass or fluid which has a very decided and thermic influence. Numerous experiments have been tried with the seeds of mignonette, many varieties of the flowering pea, the common parsley, and cresses under the various tints of glass—with all of them the seeds have germinated, but except *under the blue glass* these plants have all been marked by the extraordinary length to which the stems of the cotyledons have grown, and by *the entire absence of the plumula*—no true leaves forming, the cotyledons soon perish and the plant dies; *under the blue glass* alone has the process gone on healthfully to the end."

"The changes which take place in the seed during the process of germination have been investigated by Saussure: oxygen gas is consumed and carbonic acid is evolved; and the volume of the latter is exactly equal to the volume of the former. The grain weighs less after germination than it did before; the loss of weight varying from one-third to one-fifth. This loss of course depends on the combination of its carbon with the oxygen absorbed, which is evolved as carbonic acid."

"For the discovery that oxygen gas is exhaled from the leaves of plants during the daytime, we are indebted to Dr. Priestley; and Seunebier first pointed out that carbonic acid is required for the disengagement of the oxygen in this process. M. Theodore de Saussure and De Candolle fully established this fact."

The experiments of Seunebier show that the most refrangible of the solar rays, viz., the violet, are the most active in determining the decomposition of carbonic acid by plants.

"We have now certain knowledge. We know that all the carbon which forms the masses of the magnificent trees of the forests and of the herbs of the fields has been supplied from the atmosphere, to which it has been given by the functions of animal life and the necessities of animal existence. Man and the whole of the animal kingdom require and take from the atmosphere its oxygen for their support. It is this which maintains the spark of life, and the product of this combustion is carbonic acid, which is thrown off as waste material, and which deteriorates the air. The vegetable kingdom, however, drinks this noxious vapor; it appropriates one of the elements of this gas—carbon—and the other—oxygen—is liberated again to perform its services to the animal world."

"The animal kingdom is constantly producing carbonic acid, water in the state of vapor, nitrogen, and in combination with hydrogen, ammonia. The vegetable kingdom continually consumes ammonia, nitrogen, water, and carbonic acid. The one is constantly pouring into the air what the other is as constantly drawing from it, and thus is the equilibrium of the elements maintained."

"Beccaria examined the solar phosphori, and ascertained that the violet ray was the most energetic, and the red ray the least so, in exciting phosphorescence in certain bodies."

"M. Biot and the elder Becquerel have proved that the slightest electrical disturbance is sufficient to produce these phosphorescent effects. May we not therefore regard the action of the most refrangible rays, viz., the violet, as analogous to that of the electric disturbance? May not electricity itself be but a development of this mysterious solar emanation?"

It has been long known to chemists that a mixture of chlorine and hydrogen gases might be preserved in darkness without combining for some time, but that exposure to diffused day light gradually occasioned their combination, and which is effected with the greatest speed by the *extreme blue and indigo rays.* M. Edmond Becquerel in 1839 first called attention to the "electricity developed during the chemical action excited by solar agency."

The experiments of Dr. Morichini, repeated by MM. Carpa and Ridolfi, that violet rays magnetized a small needle, were successfully confirmed by Mrs. Somerville.

"Light is not solely a radiant visible element. It has other properties which cannot be overlooked. It oxidizes, colors, bleaches. Light becomes absorbed—light changes into heat, and heat into electricity; in fact, light in its radiant visible character only shows one of its many phases. Light holds many forces within its beams. It has properties, powers of its own, which neither mathematician nor optician can grasp. It is a great chemical agent. Colors are produced by a change resulting from a polaric act of arrestation—yellow and red yellow belong to the acids; blue and red blue to the alkalies. The undulatory theory explains the radiant visible property of light, but it does not explain its chemical effects, the optical polarity of a chrystal and its connection with the polaric condition of its constituents—the diffraction, inflection, interferences, the oxidation of surfaces as the cause of natural colors, the presence of the chemical action of light, the presence of heat, electricity, magnetism; yet light produces all these phenomena; it vitalizes, and the organic action of light is witnessed in the fauna and flora around."

We have seen that blue light, and the violet ray which is a compound of it, and the red ray—being the most refrangible rays of the solar spectrum—excite magnetism,—and electricity, by which carbonic acid gas evaporated from growing plants is decomposed and oxygen thereof liberated to be absorbed again in maturing the flowers, fruit and seed of the plant, thus stimulating the active energies of the plant into its fullest and most complete development. Now this is just what I think is done in the vegetable world by the blue light of the firmanent. That blue light of the firmanent, if not itself electro-magnetism, evolves those forces which compose it in our atmosphere, and applying them at the season, viz., the early spring, when the sky is bluest, stimulates, after the torpor of winter, the active energies of the vegetable kingdom, by the decomposition of its carbonic acid gas—supplying carbon for the plants and oxygen to mature it, and to complete its mission.

In the experiment which I have made in the cultivation of grapes under violet light, I have endeavored to combine with it the blue light of the firmanent, causing the other rays of the solar spectrum to be absorbed while the blue and violet rays were permitted to permeate the violet glass into the grapery. The difference of temperature under the white glass and under the violet glass of the grapery is supposed to have excited currents of electricity sufficient to decompose more rapidly the carbonic acid gas that had been evaporated from the leaves of the vines, than would have been done under the influence of the sunshine alone—thus stimulating the increased absorption of oxygen, and the deposit of carbon in the vines, and constantly and quickly renewing the evaporation of carbonic acid gas. The result has been seen in the wonderfully large product of fruit, accompanied by a prodigious formation of new wood, to yield the crop of fruit for the ensuing year.

The investigations that have been made during the present century regarding light have developed the existence of some remarkable attributes; one of the most astonishing is the discovery that there is no heat *per se* in the sun's ray, though it is one of the causes which produce heat. This is established beyond dispute by the existence of the intense cold which prevails in the upper atmosphere, increasing with its altitude, and through which all the sunlight which reaches the earth must pass, but whose temperature it cannot alter. Hence you have at the present time the line of perpetual snow, according to Professor Agassiz, at an elevation of 15,000 feet at the equator, of 6,000 feet at the latitude of 45°, and gradually approaching the surface of the earth till it reaches it at 60° of north latitude, beyond which ice prevails nearly to the pole.

Aeronauts have remarked also at great altitudes above the earth that the thermometer had ceased to mark any variation of temperature when exposed in the full sunshine or in shadow.

A curious illustration of the fact that something more is needed than sunlight to produce heat is to be found in the fact stated by the famous arctic navigator, Dr. Scoresby, as well as by others, that when, after a long night in the arctic regions, the sun had appeared, though the thermometer was below 32° of Fahrenheit, and everything around was frozen hard, he observed that the pitch with which the seams of the planks of the ship had been payed, on the side of the ship exposed to the sun, was melted, notwithstanding the great declination of the sun and the small angle of incidence, that the nearly horizontal rays of it made as they fell upon the pitch, while that in the shade on the other side of the ship was so hard that it was with difficulty broken with a hatchet—other objects on the ship manifesting at the same time the low temperature marked by the thermometer. I am not aware that any explanation of this phenomenon has ever been attempted. I may, therefore, offer to suggest that the pitch being an electric or non-conductor of electricity and negatively electrified when the sun's ray positively electrified fell upon it, an explosion took place, heat was evolved, and the pitch was melted—thus proving that heat from sunshine is produced by the contact of an electricity opposed to that of the sun's rays.

As a corollary from what has just been stated, it may be observed that the heat of the equatorial and tropical oceans is not derived from the sun. We do not heat our houses by kindling fires at the tops of our chimneys or boil our water from above, but rather we descend into our cellars, and make our fires for that purpose in the furnaces constructed there. Besides, we know that from the surface of the water, if at rest, and from its many surfaces, if agitated by winds, the rays of the sun would be reflected in all possible angles corresponding to the angles of incidence of the rays themselves, and the heat would be lost in space. Whence comes, then, this ocean heat in the tropics, finding its vent in the arctic and antarctic regions through the Gulf Stream of the Atlantic, and the Japan Stream laving the shores of northeastern Asia, and the south-eastern current running along the south-western coast of South America to the Antartic seas? Does it not come by radiation from the interior of the earth from those great fires which, by the elastic gases and vapors engendered there, in many parts of the world upheave mountains and islands, and forming chimneys for themselves in their summits, belch out that superfluous heat, light, electricity, and magnetism which radiation to the surface of the earth at times is inadequate to discharge? And are not these great ocean currents of heated water merely channels or flues of radiation of heat from beneath, by which, for climatic purposes, the Omnipotent Creator has devised the means of distributing this interior heat over the surface of our planet?

All admit the existence of those great forces of nature in the interior of the earth, manifested through volcanic action in those imponderable elements of heat, light, electricity, and magnetism. Why are those forces there? May they not be the forces which turn the earth on its axis, and aid in propelling it around the sun? May not the frigid zones north and south furnish the cold cushions of water in the extreme depths of the ocean, of the uniform temperature of 39½° of Fahrenheit, and of nearly the greatest density known to that element, for the purpose of restraining and controlling the radiation of that great interior heat of the earth, which otherwise might be wasted?

Dr. Winslow, in his treatise on light, its influence on life and health, says: "Accurate calculations have been made as to the temperature of the ocean. The results obtained clearly establish that the lowest degrees of temperature are obtainable on the surface of the water; and that about ten feet below the surface the thermometer rises several degrees,—90° is said by Mr. Agassiz (son of Professor Agassiz,) to be the highest temperature he has known the ocean to attain; at very great depths of the ocean a uniform temperature of about 39½° has been found."

The low temperature of the surface water of the ocean is attributable to the evaporation which is constantly going on, carrying off the atmospheric heat adjacent, and proving conclusively that the Gulf and other warm ocean currents do not derive their heat from the sun.

These reflections have forced themselves upon me, while pondering over some of the great revelations of nature.

In a recent report of the Secretary of the Agricultural Bureau at Washington, he states—"On the 15th of June the sun is more than 23° north of the equator, and therefore it might be inferred that the intensity of heat should be greater at this latitude than at the equator; but that it should continue to increase *beyond this even to the pole*, may not at first sight appear so clear. It will, however, be understood when it is recollected that though in a northern latitude the obliquity of the ray is greater, and on this account the intensity should be less, yet the longer duration of the day is more than sufficient to compensate for this effect and produce the result exhibited."

It strikes me that this explanation is not sound. I remember several years ago, at Philadelphia, on the afternoon of a day in August, when the thermometer was at 94°, that in fifteen minutes the thermometer fell 40°, which was owing no doubt to a descending column of cold air from the upper atmosphere, attracted by some local electrical disturbance. The continuous heat of the preceding summer months could no more prevent this thermal change at Philadelphia than could the long day with the oblique sun's rays increase the intensity of the heat in high northern latitudes.

Professor Maury says—"The summer temperature as observed on the very borders of the Polar ocean is absolutely marvelous. Observations made with a view of determining this accurately have for some years been taken in Alaska. One of the observers in the northern district of Yukon states in the 'Agricultural Report' for 1868, 'I have seen the thermometer at noon at Fort Yukon, not in the direct rays of the sun, standing at 112°; and I am informed by the commander of the post that several spirit thermometers graduated to 120° had burst under the scorching sun of the arctic midsummer, which can only be appreciated by one who has endured it. In midsummer, on the Upper Yukon, the only relief from the intense heat under which vegetation attains an almost tropical luxuriance, is the two or three hours during which the sun hovers near the northern horizon, and the weary voyager in his canoe blesses the transient coolness of the midnight air.' "

According to M. de Humboldt, the sky is bluer between the tropics than in the higher temperate latitudes, but paler at sea than in the interior of countries; the blue is less intense at the horizon than at the zenith. The early maturity of human life in the tropics is to be attributed to the stimulating influence of the enormous quantities of electricity, which, continually passing by day as well as by night in the auroras from the poles to the equator, and descending to the earth in those regions, in those dazzling sheets of lightning flame, so terrifying to all who have witnessed them, and conducted by the incessant rains prevailing there in certain seasons of the year—deoxygenate the enormous volumes of carbonic acid gas generated by the exuberaut vegetation, as well in its growth as in its decay, thus supplying excessive quantities of oxygen gas to stimulate and support the animal life, as well as carbon to the fresh vegetation which is being continually renewed—the circle of development and decay in the vegetable kingdom being thus always preserved.

We have thus seen that the magnetic, electric, and thermic powers of the Sun's ray reside in the violet ray, which is a compound of the blue and red rays. These constitute what are termed the chemical powers of the sunlight. That they are the most important powers of nature, there can be no doubt, as without them life cannot exist on this planet. Without these chemical powers there could be no vegetation. Without vegetation there could be no insect life, and no development of the higher order of animal existence. The earth would be without form and void, and we can now understand the potential meaning of the first sublime utterance of the Almighty in forming this earth, when he said "Let there be Light," and there was Light.

From the foregoing premises we deduce the following conclusions:

1. Heat is developed by opposite electricities in coujunction and in proportion to the quantity and intensity of those electricities in contact with each other, will be the intensity of the heat.

2. The blue color of the sky, for one of its functions, deoxygenates carbonic acid gas, supplying carbon to vegetation and sustaining both vegetable and animal life with its oxygen.

www.ingramcontent.com/pod-product-compliance
Lightning Source LLC
LaVergne TN
LVHW011135110826
845150LV00008B/2363
* 9 7 8 1 4 1 8 1 9 3 7 4 4 *